Engineered Poisxons Coronaviridae Coronavirus

By Mister. E Tyler Lazarus Stump

Engineered poisxons:
AKA
Coronaviridae CORONAVIRUS

By Mister. E

This is a pre-origin book to explain how a novel coronavirus
Was created and later got out; pertaining to the deadly 2019/2020
Pandemic which has now changed the entire world.

Tyler Lazarus Stump

ISBN

Printed, Distributed and Bound in the United States of America First Printing April 2022

Published by He Who Rebels Against All

Oklahoma City, Oklahoma 73106

Hey, Thanks for getting a copy!

Facebook.com/TylerLazarus1992

 Facebook.com/Oddityler

Facebook.com/KFPSeries

We know the pathogen outbreak
epicenter came from Wuhan.

We know the pathogen came out of the
Wuhan Institute of Virology.

We know this virus came from an
institute of virology.

We also know the pathogen was
researched in that lab,
for years, before it was released.

Accidentally, but likely purposefully.

We know After the first SARS (SARS1)
outbreak, mainland China and all of its
BSL-3, BSL-4 level labs expressed a
natural interest in researching the
pathogen to ensure another never got
out.

THAT FAILED.

AND SOME OF THEM EXPRESSED A NEGATIVE
CURIOSITY IN THE PATHOGEN.

WE KNOW THIS LAB WAS RESEARCHING AND
PERFORMING WORK TO ENHANCE THE
VIRULENCE OF CORONAVIRUSES,
AND THEY HELD A SAMPLE WITH A 98.6%
GENETIC MATCH TO THE MICHAEL MYERS
VIRUS THEY LET OUT.

WE ALSO KNOW CHINESE MILITARY
BIODEFENSE WORK, THROUGH LEAKED AND
CLASSIFIED DOCUMENTS,
HAVE BEEN WORKING ON MANIPULATING/
CREATING NEW VIRUSES THAT CAN TARGET
SPECIFIC RACIAL AND ETHNIC GROUPS.

ALEXA, PLAY "TOXIC" BY BRITNEY JEAN
CORONA SPEARS.

WITH THE COVID PARTICLES ON YOUR LIPS,
I'M ON AN ANTIBODY HIGH

Your toxic respiratory pathogens
slipping under

CHAPTER 2

While you're busy worried about
offending the chinese, their military
sectors are working on creating
viruses to kill you.

And other people, gentically like you.

Lmao.

bangs Gong

chopsticks and non-working vaccine
technology in the mix,
the biotechnology panda gets his kicks

While RUssia is at war with Ukraine,
And seeding disinformation
throughout the United States and the
world,

Xhina is in the lab, screwing with the
human genome,
and splicing chimeras together.

Oh, what a life.

Who needs enemies with "friends" like
that?

But it wont matter. Xhina wont be
stopped, and neither will Rush-uh.

The world has gone soft, complacent,
stupified by the trauma and chaos
unleashed.

There are no churchills here,
only molehills, hollywood hills and
cowards.

and the truth is,

ALL OF THIS, EVEN THE MILLIONS DEAD.

IT'S CHILD'S PLAY FOR RUSSIAN AND CHINESE AND AMERICAN CHILDREN.

MORE DEADLY PATHOGENS WILL BECOME CREATED.

10 MILLION IS NOTHING COMPARED TO THE BILLIONS, THE TRUE *UNSTOPPABLE, HELL PLAGUE* MADE IN A LAB WILL CREATE.

I'M SURE THEY'RE WORKING ON IT NOW.

TANKS, NUKES, AIRCRAFT, MISSILES, GUNS.

ALL OLD SCHOOL.

EXPENSIVE. LABOR INTENSIVE TO CREATE.

WHY NOT JUST SEED AN INVISIBLE KILLER OUT INTO THE MORNING AIR.

TAKES LESS MONEY, LESS TIME.

LESS DRAMA.

BUT MORE IMPACTFUL.

POISONS, ATTENUATED PATHOGENS,
ENGINEERED VIRUSES ARE THE FUTURE.

BIOWEAPONS WILL PLACE COMBAT WEAPONS.

BIOWARFARE WILL REPLACE NUCLEAR
WARFARE.

WITH WARFARE, WEALTH IS DESTROYED.

BOTH PERSONAL, GLOBAL AND THE
INFRASTRUCTURE FINANCIALS.

A VIRUS ENSURES ALL OF THAT REMAINS IN
PLACE,
WHILE THE PEOPLE HOLDING THAT WEALTH,
DIE.

SIMPLE AND CLEAN.

HIGHLY PROFITABLE, TOO!

What you've just seen for the past few years, is an orchestrated play.

A theatre stage with many bad actors, and several acts.

It's the playbook
for things to come.

Nukes, thermnuclear and thermobaric tech all certainly provides that Movie, cinematic, fiery climax.

But virual specimens are and will be the way men seeking power achieve their ends.

I'm not saying nukes and increasingly destructive firepower wont exist.

But biowarfare will eclipse the traditional world war 1,2 and 3 warfare.

Simpler to use, simpler to make, simpler to store and keep.

*simple, being relative in scientific terms.

Enriching uranium and keeping it safe is of an order more complicated than splicing a nucleic acid sequence into a piece of simple genetic material.

Literally.

designer bio-weapons.

For 2, annoying, Migraine inducing years, i had lab notes, leak leaks, and lab journals, now deleted lab papers but no solid lab connection.

I had all the players, but I was still being played.

I knew Fauci had funded the wuhan institute.

I knew Shi Zhengli had studied coronaviruses at that institute.

But I didnt have a link of who or how they had been studied, previously, outside of the sars1 outbreak.

until I realized, I actually had the missing piece and even a photo.

It had *literally* been looking at me.

Contrary to popular opinion, The Fault of COVID19 isnt all On China.

Of course, they were creating bioweapons to kill and infect, of course they covered up covid19, of course they knew it was circulating, before they told anyone.

For almost 3 months of ... Insane

infection... I'm sure.

But The creation of that virus was helped, backed, and performed by United STates corrupt money and science.

Fauci funding a lab, and set of chinese scientists studying coronavirus and increasing virulence wasn't enough.

How the hell did this biohazard became so deadly between sars-1 and sars-2.

Naturally, I knew it hadn't evolved or mutated.

It hadn't become 300x more deadly from nature.

I couldn't understand the final, missing connection.

I knew who made it more deadly.

But.... how.

How, indeed.

Our story goes back to 2012.

A group of workers were in an abandoned copper mine, and came into contact with bat guano (bat poop) inside of this cave network.

They touched it, and immediately became sick with a mystery illness. Within a week.

Several of them died.

Doctors in China couldn't figure out what illness they had. The workers continued to exhibit respiratory

pathology and airway issues as they clung to life.

They tested them for everything.

Influenza. Hemorrhagic fever.

You name it.

Nothing.

Until they tested them for Sars.

Instant match.

Now.

Not only does this lead to the scientific body - as an entire institution - now finding the missing link to sars1.

They knew it had jumped into civet populations but couldn't locate the ancestor match of where that civet became originally infected.

So many things are about to fall, perfectly, into place.

Try and keep up.

This final, definite proof of where sars came from, manifests in this test.

The remaining workers that beat this sars infection, leave the mine behind.

It, however, remains open.

And a certain Shi-Zhengli, dubbed at batwoman, takes several trips out to this infamous mine.

One year later and collects a shitload (literally) of bat shit samples.

And i believe some of the wuhan institute workers who collected those samples, also became ill.

But it's unconfirmed because the corresponding patient logs are all

Written in Chinese and I can't authenticate the source and half of it was redacted and blacked out.

That being said, she most definitely took a trip out there and collected samples.

I've got the screenshot now, of the pubmed data she logged. Years ago.

Check this shit out:

Virus name: hCoV-19/bat/Yunnan/RaTG13/2013
Acession id: EPI_ISL_402131
Type: Sars related Coronavirus
Passage details/history: original sample

Sample Information: 2013-07-24
Location: Asia/ China/ Yunnan/ Pu'er
Host: Rhinolophus affinis
Gender of sample: male
Patient age: unknown
Specimen source: Fecal swab

Sequencing technology: illumina hiseq
3000
assembly method: CLG genomics
workbench v12.0

originating lab: wuhan institute of
virology, chinese academy of sciences
address: 44 Xiao hong shan, wuhan
Hubei 430071
Submitter: Zhengli Shi

THE SMOKING BAT GUN, OF THE SMOKING BAT
SHIT.

GET TO THE BATMOBILE, BATMAN.

hcov-19/bat/yunnan/ratg13/2013 is so
important here.

RATg13 is one of the samples with a 98%
genomic match to covid 19.

This is Sars, in a glass Biocontainment
bottle.

THE ORIGINAL

THE O.G.

THE ORIGINAL GANGSTER OF COVID19.

NOW.

INTERESTINGLY ENOUGH, SHI ZHENGLI
DOESNT LOG THIS AND AUTHENTICATE IT FOR
YEARS.

SHE HAS THE SAMPLE, AND HAS NOW LEFT AN
ORIGINAL SOURCE DOCUMENT/PAPER TRAIL.

BUT THERE'S NO DATA LOG OF IT.

SHE BEGINS TO SEQUENCE AND SAMPLE IT,

AND FROM HERE, WE HAVE THE BACKBONE OF
THE VIRUS THAT WILL TAKE SARS1 TO BECOME
SARS2.

YOU HAVE A SARS CAUSING AGENT, AND NOW

YOU HAVE THE ABILITY TO AMPLIFY IT, AND MAKE IT JUMP HOST SPECIES.

AND THAT'S EXACTLY WHAT THEY DID.

BUT WAIT, WHO WAS SHE WORKING WITH.

IT'S NOT FAUCI.

HE'S JUST THE GRANT FUNDING, COIN PURSE, POLITICAL SCIENCE ADMIN AGENT TO IT ALL.

OF COURSE HE ALLOWED AND WANTED IT TO SPREAD, AS EVIDENCED BY HIM TELLING THE WHITE HOUSE TO NOT DO FLIGHT SHUTDOWNS OUT OF CHINA, ALLOWING INTERNATIONAL TRAVEL TO REMAIN OPEN, AS CHINA ITSELF WAS PATIENTLY WAITING FOR IT TO SPREAD.

*THIS WAS A WARGAME, I HOPE THAT'S CLEAR, BY NOW.

BUT FAUCI WAS NOT THE SCIENTIST IN THE

LAB, OR THE SCIENTIST WORKING ON THE SPECIMEN ITSELF.

SHI NEEDED ANOTHER.

AND NOW ENTER, R. BARIC.

I HAD PHOTOS OF THE GUY IN MY NOTES, AND DIDNT REALIZE IT WAS THATTT GUY.

I LAUGH NOW, BECAUSE I FLIPPANTLY THOUGHT HE WAS JUST ANOTHER "RESEARCHER."

OH, NO.

R. BARIC IS THE FATHER OF CORONAVIRUS RESEARCH.

IF FAUCI IS THE EVIL GODFATHER OF GAIN OF FUNCTION AND PATHOGEN ENHANCEMENT,

BARIC IS THE KING OF KUHRONA.

Shi-Zhengli is smart. Evil and corrupt.

But smart.

But intelligence of one isn't enough to amplify this pathogen in all of the ways it was.

I needed the link of who and what was experimenting around on it, before it escaped/was released.

Barric and Shi, happen to have multiple papers, published together.

I have all of them.

I kept looking for her name, without realizing, his name is tied in almost *all* of them.

Especially all of the major ones, experimenting around with pathogenic virility.

Specifcally with Sars, MERS and Sars-Coronaviruses.

After all, he was one of the first to respond to the first Sars outbreak in 2002.

This was the link.

I had Fauci.
I had Shi.

But who was the american helping Shi.

It was he.

Now, for the sake of journalistic integrity, I'm not making any claims about him or this man.

I dont know whether he had evil intent.

I do know he experimented specifically with causing it to infect human airway cells.

I do know he published papers notating how deadly it was at Chapel Hill, NC university.

I do know he recognized - as they made it more dangerous - it was incredibly resistant to all known medical technology.

I'm not saying anything beyond that.

But he most definitely helped make it what it IS.

He also received a material exchange request from Fauci/NIH/ and Moderna for a coronavirus vaccine candidate in 2019, months before the pandemic even started. Months before anyone even knew Covid19 existed.

It seems odd to have "the cure," before

THE POISON EVEN (PUBLICALLY) EXISTS.

I WONT SAY MORE.

BUT I WILL SAY, THERE ARE LIES AND LIARS IN
EVERY DIRECTION,

CLOAKED IN LAB COATS, AND
MASQUERADING THEIR INTENTIONS AND
PSYCHOPATHIC MOTIVES BEHIND "$CIENCE."

I KNOW THERE ARE ENGINEERED POISXONS.

WHAT I DONT KNOW IS WHY THEY HAVEN'T
BEEN HELD RESPONSIBLE

FOR POISONING THE WORLD.

THE 4 HORSEMEN HAVE STARTED TO RIDE.

RIDE OR DIE,
BECAUSE DEATH IS NOW HERE.